Public Schools
Library Protection Act 1998

Why Can't I Fly?
and other questions about the motor system

by

Sharon Cromwell

Photographs by

Richard Smolinski, Jr.

Series Consultant

Dan Hogan

RIGBY INTERACTIVE LIBRARY
DES PLAINES, ILLINOIS

© 1998 Reed Educational & Professional Publishing
Published by Rigby Interactive Library,
an imprint of Reed Educational & Professional Publishing,
1350 East Touhy Avenue, Suite 240 West
Des Plaines, IL 60018

02 01 00 99
10 9 8 7 6 5 4 3 2

Printed in Hong Kong by Wing King Tong

Library of Congress Cataloging-in-Publication Data

Cromwell , Sharon , 1947-
    Why can't I fly? : and other questions about the motor system / Sharon Cromwell ; photographs by Richard Smolinski, Jr.
    p.   cm. -- (Body wise)
    Includes bibliographical references and index.
    Summary: Introduces the human motor system and how it allows the body to move, covering such aspects as bones, muscles, balance, and exercise.
    ISBN 1-57572-159-7 (lib. bdg.)
    1. Musculoskeletal system--Juvenile literature.   [1. Human locomotion.
2. Muscular system. 3. Skeleton.] I. Smolinski, Dick, ill. II. Title. III. Series.
QP301.C69   1997
612.7--dc21   97-22227
                                                                        CIP
                                                                        AC

Some words are shown in bold, **like this.** You can find out what they mean by looking in the glossary.

Contents

What is my motor system?

Your body moves with the help of your motor system. Your muscles and bones are part of this system. Your muscles are attached to your bones. You have more than 600 muscles in your motor system. When your muscles squeeze in and press out, they move your bones. That makes your body move.

Another system, called your nervous system, tells your muscles when to squeeze in and when to press out. When a muscle squeezes in, it contracts. When a muscle presses out, it extends.

HEALTH FACT
Vitamin D is a substance that is good for your bones. You can get a healthy dose of vitamin D by playing outside on a sunny day.

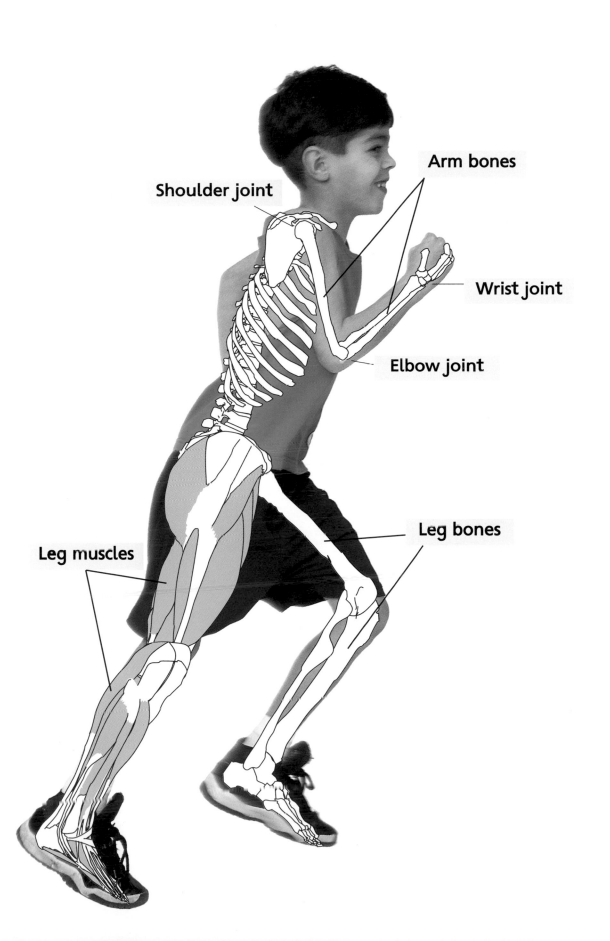

Arm bones

Shoulder joint

Wrist joint

Elbow joint

Leg bones

Leg muscles

How does my motor system help me move?

Your skeleton is made up of all of your bones. Your skeleton is an important part of your motor system. Your **joints** are also part of your motor system. A joint is a place where two bones meet. Your motor system works well if your joints move easily and your muscles and bones are strong.

There are three kinds of joints in your motor system. Your shoulder has one kind of joint. It moves in all directions. Your elbow has another kind of joint, which only moves backward and forward. Your wrist has a third kind of joint, which twists so you can turn your hand over. Joints that move are padded by a rubbery substance called **cartilage**. Your bones are held together by **ligaments.** When you kick a soccer ball, you use many bones, muscles, and joints!

HEALTH FACT

Bread, cereal, and pasta are good foods to eat when you need a lot of energy for playing a sport.

Why can't I fly?

Your body can't leave the ground on its own because of a force called **gravity** and because of your body's shape.

PhotoDisc, Inc.

Gravity

Arm
muscles

• Birds fly because their wings are very large compared with the rest of their bodies—which are light.

• Birds are lifted by flapping large, powerful wings. Their bodies are light enough for their wings to lift them.

• Gravity is a force that keeps everything on the ground. It pulls you down and keeps you from flying off into space.

• Your arms are small compared with the rest of your body. They aren't wide enough to lift you off the ground.

Why do I have more bones than a grown-up?

As you get older, some of your bones grow together. You have fewer and fewer bones as you grow up, but your bones get longer.

HEALTH FACT

Calcium is one of the most important nutrients for healthy bones. Milk and dairy products have a lot of calcium.

1. A baby's bones start to grow while the baby is still in the mother's **womb**.

2. You had 350 soft bones when you were a baby.

3. As you get older, two bones grow together to become one long bone. This happens to only some of your bones.

4. That's how you end up with 206 bones when you are an adult.

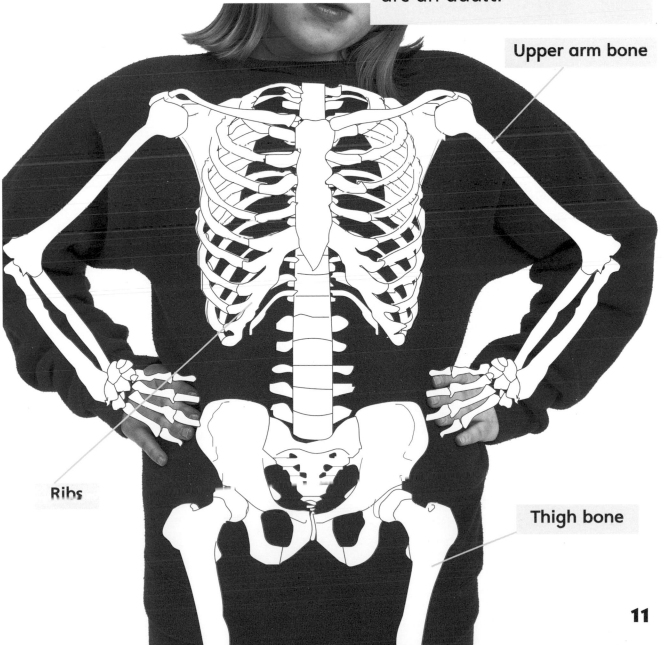

Upper arm bone

Ribs

Thigh bone

Why do I need to "warm up" my muscles before exercise?

Your muscles work better—stretch better—when they are "warm." Using them gently at first gets the blood flowing. This "warms" up the muscles.

HEALTH FACT

To warm up the muscles in your arms and legs, try some gentle stretching exercises.

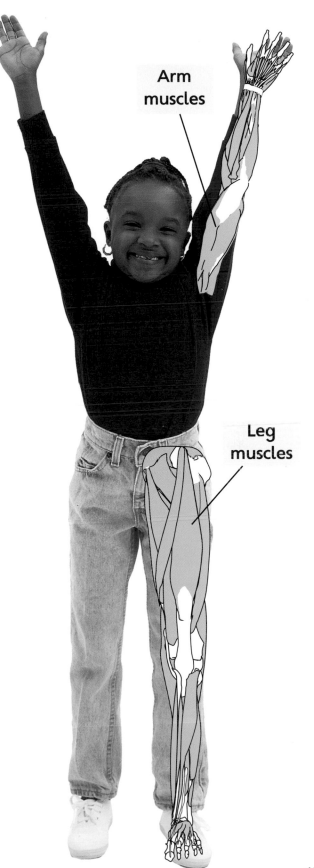

Arm
muscles

Leg
muscles

● During exercise, your
muscles contract and
extend very fast.

● If you haven't warmed
them up, they will not be
elastic, or stretch well.

● When muscles are not
elastic, they may strain or
even tear.

● Strained or torn muscles
will hurt. These muscles
need time to heal.

How do I balance to ride my bike?

Your brain learns which muscles need to contract and extend in order for you to move in certain ways. Your brain then sends messages to your muscles telling them to move in that way.

HEALTH FACT

Exercise that makes you breathe hard, such as an uphill ride on a bicycle, is good for your heart and your lungs.

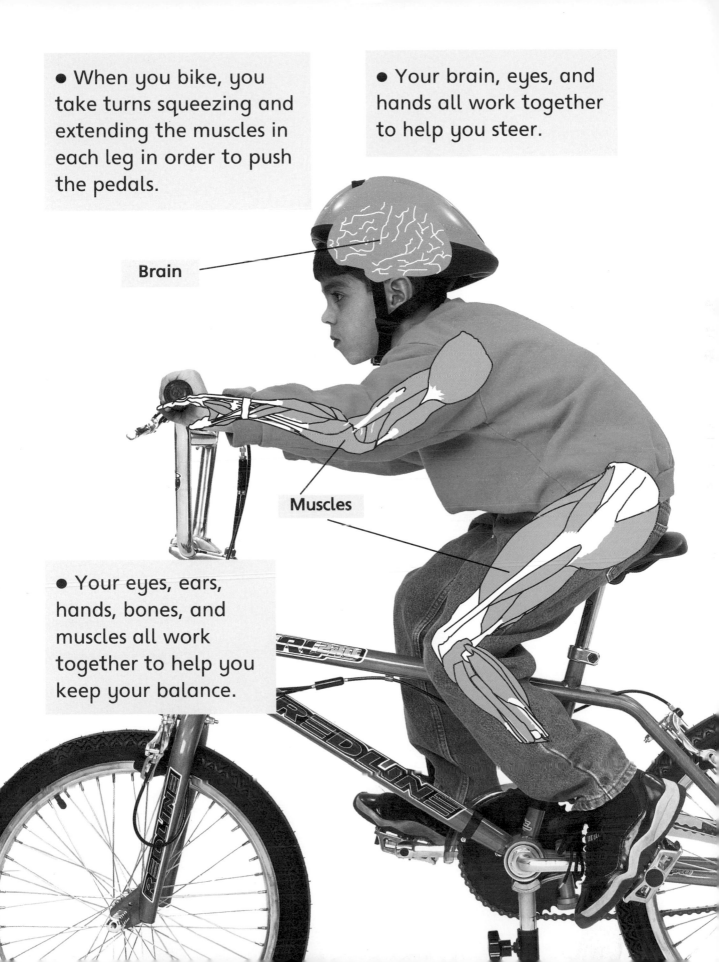

● When you bike, you take turns squeezing and extending the muscles in each leg in order to push the pedals.

● Your brain, eyes, and hands all work together to help you steer.

Brain

Muscles

● Your eyes, ears, hands, bones, and muscles all work together to help you keep your balance.

Can exercise ever be harmful to me?

You can get hurt when you exercise or play a sport.

HEALTH FACT

When you exercise or play a sport, use proper equipment, such as weights that are right for you. This will help reduce the chance of an injury.

- One type of injury from sports is called a **stress fracture**.

- Stress fractures are thin cracks or weaknesses in the bone.

- Exercise or sports that involve a lot of running and jumping can sometimes cause stress fractures.

- If you feel pain while exercising, stop right away.

- If you keep exercising or playing, you could get a stress fracture.

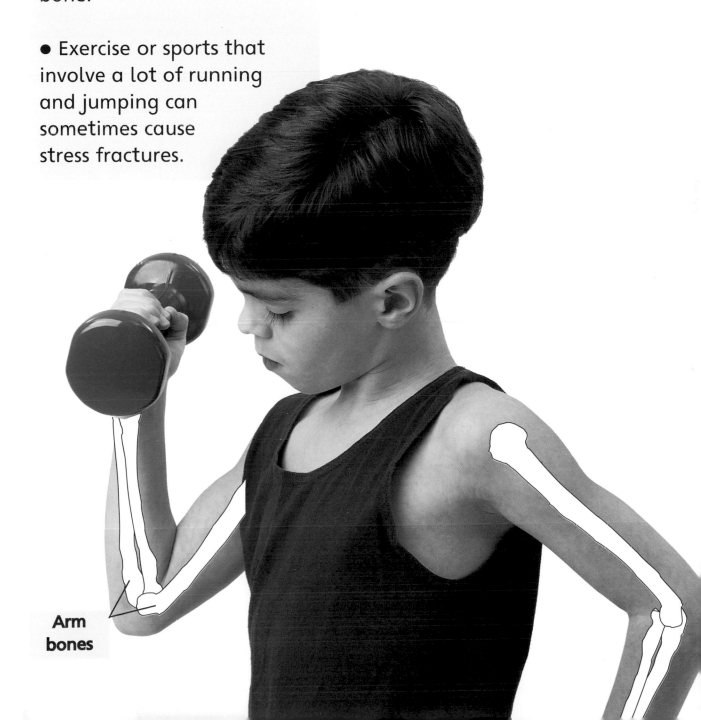

Arm bones

Why can I do so many things with my hands?

Your hands are one of the most amazing parts of your motor system. With so many bones and muscles in one tiny area, your hands are able to do a great variety of things.

HEALTH FACT

Many arts and crafts activities—such as drawing—help you become better at using the muscles in your hands.

- You have 27 bones in each hand.

- The large number of bones and muscles in your hands makes them very strong, but they can also be very gentle.

- Your thumb also allows you to do many things with your hands.

- When your thumb and fingers work together, they can do everything from lifting heavy objects to holding onto a pencil.

- No other part of your body is able to be so strong and so gentle at the same time. Your hands are also some of the most sensitive parts of your body.

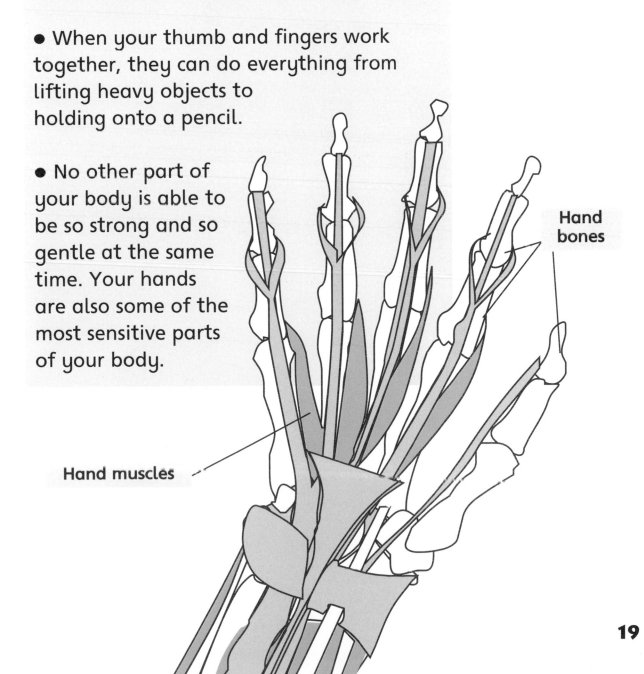

Hand bones

Hand muscles

Why is exercise good for me?

Exercise helps make your body stronger and helps make you feel better.

HEALTH FACT
Doing pull-ups makes your arm muscles and your heart stronger.

● When you breathe more during exercise, your heart pumps harder. This makes the blood carrying oxygen move faster through your body.

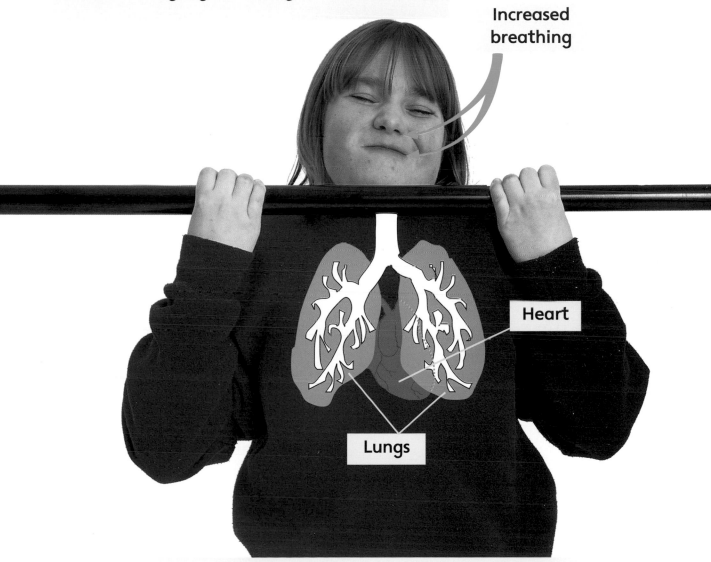

Increased breathing

Heart

Lungs

● When your muscles work hard during exercise, they get bigger and stronger. This also makes your body stronger.

● Your heart is a muscle too. The more it is exercised, the healthier it becomes.

EXPLORE MORE!
Your Motor System

1. GO AHEAD AND JUMP!

WHAT YOU'LL NEED:

• a friend to help you

THEN TRY THIS!

Jump as high as you can without bending your knees. Now ask your friend to bend her knees and jump. Who jumped higher? When your knees are straight, the only muscles that can contract are the short ones in your feet. When your knees are bent, you can contract the long muscles in your legs. The greater the contraction you make, the higher you can jump.

2. TAKE A STAND!

WHAT YOU'LL NEED:

• a friend to help you

THEN TRY THIS!

Stand on one foot and hop with your arms out to help you balance. Have a friend count how many times you can hop. Then close your eyes, put your arms down, and try it. Can you hop as many times this way? Without your eyes and arms, your sense of balance is very different!

First, bounce the ball on the ground. How long can you keep it bouncing? Have a friend time you. Then have your friend time how long it takes you to pick up 20 toothpicks, one at a time. Now you time your friend. Who had the better large muscle control? Whose small muscle control seemed better?

3 CONTROL YOURSELF!

WHAT YOU'LL NEED:

- 20 toothpicks
- a ball to bounce
- a watch or sports watch
- a friend to help you

THEN TRY THIS!

There are basically two kinds of muscle activities that your body can do. There are large muscle activities, such as swinging a baseball bat and running. Then there are small muscle activities, such as writing with a pen or buttoning your shirt. This activity will show you about large and small muscle control.

23

Glossary

cartilage Rubbery padding for the bones.

gravity Force that pulls things down toward the earth's surface.

joint Place where two bones meet.

ligament Material like string that holds bones together.

stress fracture Crack or weakness in the bone.

womb The part of a woman's body in which a baby grows.

More Books to Read

The Amazing Body. Philadelphia, PA: Running Press, 1994.

Bryan, Jenny. *Movement: The Muscular and Skeletal System.* Morristown, NJ: Silver Burdett Press, 1993.

Elting, Mar. *Book of the Human Body.* New York: Simon and Schuster Childrens, 1986.

Kalman, Bobbie. *My Busy Body.* New York: Crabtree Publishing Company, 1985.

Rice, Christopher and Melanie Rice. *My First Body Book.* New York: Dorling Kindersley, 1995.

Index